EMPLOYEE'S RECEIPT

I acknowledge receipt of Keller's *Food Safety Handbook for Foodservice Employees*. The handbook covers the following nine topics regarding safety in foodservice and a glossary:

Preventing Hazards in Food

How to Limit Cross Contamination

Personal Hygiene

Time and Temperature Controls

Sanitation

HACCP

Cuts and Knife Handling

Slips, Trips, and Falls

Lifting Techniques to Save Your Back

Glossary

____

Employee's Signature   Date

____

Company Name

____

Company Supervisor's Signature

> NOTE: This receipt shall be read and signed by the employee. A responsible company supervisor shall countersign the receipt and place in the employee's training file.

Food Safety Handbook for Foodservice Employees

J. J. Keller & Associates, Inc.
3003 W. Breezewood Lane, P. O. Box 368
Neenah, Wisconsin 54957-0368
Phone: 1-800-327-6868
www.jjkeller.com

Library of Congress Catalog Card Number:

00-108602

ISBN 1-57943-867-9

Canadian Goods and Services Tax (GST)
Number: R123-317687

Printed in the U.S.A.

Due to the constantly changing nature of government regulations, it is impossible to guarantee absolute accuracy of the material contained herein. The Publishers and Editors, therefore, cannot assume any responsibility for omissions, errors, misprinting, or ambiguity contained within this publication and shall not be held liable in any degree for any loss or injury caused by such omission, error, misprinting, or ambiguity presented in this publication.

This publication is designed to provide reasonably accurate and authoritative information in regard to the subject matter covered. It is sold with the understanding that the publisher is not engaged in rendering legal, accounting, or other professional service. If legal advice or other expert assistance is required, the services of a competent professional person should be sought.

The Editorial Staff is available to provide information generally associated with this publication to a normal and reasonable extent, and at the option of and as a courtesy of the Publisher.

TABLE OF CONTENTS

INTRODUCTION TO THE EMPLOYEE

Your job in foodservice is important—not only to you, but also to your employer and the customers you serve. You are responsible for satisfying customers with good tasting food that is safe to eat. This means food that will not cause people who eat it to become sick.

This handbook will help you learn more about how to keep the foods you work with safe. You'll become aware of potential hazards that can make food unsafe, and learn the correct practices to prevent those hazards from taking place. Plus, you'll get to know habits to help keep you safe—from backache, falls, and accidental cuts from knives.

Your company has certain hygiene, sanitation, and other cleanliness requirements that help to ensure you serve safe food. It's important for you to know and follow all the rules your employer has set out.

This handbook provides additional information that will enable you to keep yourself, the food you work with, and those who eat it safe. Your company may use this handbook in food safety training sessions. In addition, you can keep this handbook in your locker or designated storage area at all times, to serve as a handy reference for food safety information.

Remember, your own safety and that of your co-workers and customers depends on the way you do your job. Follow your company's procedures and the guidelines in this book to do your part.

PREVENTING HAZARDS IN FOOD

As a foodservice professional, your goal, and that of your employer, is to satisfy your customers with good tasting, quality food. Your job depends on this.

But it also depends on something more. You are responsible for serving food that is safe to eat.

What does this mean? Specifically, it means food that is free of objects that don't belong there, free of bacteria that could cause illness, and free of chemicals that may be used in the area where you work.

It's everyone's job to see that none of these hazards comes into contact with the food you serve. This book will help you do that. We'll begin this chapter by defining the hazards themselves, and then go on to explain the steps you can take to be sure that the food you serve is, indeed, safe.

Three types of hazards

A hazard—or a source of danger—in food can be one of three types:

1. Physical
2. Chemical
3. Biological

In brief, a **physical hazard** is an object that can get into food but does not belong there. Physical hazards could include bits of glass or dishware, fingernails, chips of wood, or jewelry.

A **chemical hazard** is a chemical that might accidentally get into food, making the food dangerous to eat. Chemical hazards could be cleaners, sanitizers, pesticides, paint, or allergens.

A **biological hazard** in food usually refers to bacteria (we sometimes call them "germs" or "bugs") that can contaminate food and cause illness or even death. Some of the most common and dangerous biological hazards are *E. coli*, *Salmonella*, and *Staph*.

Hazards in food can be physical, chemical, or biological.

According to the Centers for Disease Control and Prevention, where outbreaks of illness or disease have been caused by mishandling of food, "most of the time the mishandling occurred within the retail segment of the food industry (restaurants, markets, schools, churches, camps, institutions, and vending locations) where ready-to-eat food is prepared and provided to the public for consumption."

You need to be aware of possible hazards and how to prevent them. Later in this book we will discuss physical hazards and chemical hazards in greater detail. Right now, we'll begin by taking a closer look at biological hazards and what you can do to keep them from making food unsafe to eat.

Biological hazards

Biological hazards include viruses, molds, parasites, and harmful bacteria. They are called "biological" hazards because they are living organisms that can grow.

These hazards are so small that they rarely can be seen with the naked eye. You need a microscope to see them. So they are also referred to as "microorganisms" (or living microscopic organisms).

Harmful microorganisms can get into food in a number of ways, and cause painful, prolonged illness if the food is eaten. For some people—particularly young children, the elderly, and those who are sick—illnesses caused by harmful microorganisms are even more serious and can be life threatening.

Dangerous microorganisms can only be seen through a microscope.

Controls to minimize bacterial growth

Your employer has probably already taken measures to prevent contamination of food from molds and bacteria. Examples of some of these measures are discussed below.

Time and temperature controls—These are an extremely effective way to control the growth of bacteria and molds in food. Keeping foods for too long in the "Temperature Danger Zone" (40°–140° Fahrenheit or 4°–60° Celsius) can accelerate growth of harmful molds and bacteria. Bacterial growth can be slowed and even stopped by certain time and temperature controls.

Moisture controls—These serve to minimize harmful bacteria growth. If a product is dry or has a low moisture level, then bacteria can't grow as easily.

Time and temperature controls help to ensure food safety.

Sanitation—If equipment is kept clean, with no food residue, it takes away bacteria's food source, and prevents them from growing.

pH—This is a measure of acidity. Bacteria do not like very acidic conditions. We preserve food by adding acid, such as in pickling.

Restriction of spreading—Preventing bacteria from moving from one place to another controls their spread and growth. (This travel of bacteria is known as cross contamination and is discussed in the "How to Limit Cross Contamination" chapter.) It is especially important to control cross contamination from raw to cooked foods, and to use good personal hygiene to prevent spreading harmful human bacteria, like *E. coli*.

The microorganisms in food that cause illness are common and potentially deadly. Your hands can pass many of them on. That's why thorough handwashing, particularly before or after certain actions, is crucial.

Harmful microorganisms

The bacteria in food that cause illness may have long, scientific names, but what they do to you is easy to understand.

Simply put, these bacteria infect people who eat the contaminated food. The bacteria can either produce the toxin (poisonous substance) in the food before the food is eaten or in a person's intestines after the food is eaten. Once infected, people can become very sick because the body tries to reject the toxin. Symptoms of infection include abdominal cramps, fever, nausea, vomiting, diarrhea, and dehydration that can lead to death.

Contaminated food can cause cramps, fever, diarrhea, even death.

Some of the more common illnesses caused by bacteria, and their symptoms, are outlined in the following table.

	Source	**Symptoms**	**Preventive Practices**
Botulism *(Clostridium botulinum)*	Improperly canned low-acid foods, or under-procesed home-prepared foods	Blurred vision Dry mouth Difficulty swallowing Difficulty breathing	Proper heating of food Maintaining low pH and water activity Temperature controls Cleaning methods
Clostridium perfringens	Slow cooling or slow reheating of food	Nausea Diarrhea Mild disease	Time and temperature controls Rapid reheating and cooling Good personal hygiene
E. coli (Escherichia coliform)	Usually contamination from contact with feces	Severe abdominal cramps Diarrhea Fever Dehydration that can lead to death	Thorough cooking of food Good personal hygiene Proper refrigeration of food Proper sanitation practices
Listeriosis *(Listeria monocytogenes)*	Unpasteurized milk, soft cheeses, eggs, and neglected, hard-to-clean spots in kitchen areas	Sudden fever Intense headache Nausea Vomiting Delirium Coma	Proper heat treatment or irradiation Preventing cross contamination Refrigeration to control growth Thorough hand washing
Salmonellosis *(Salmonella)*	Contaminated raw eggs, milks, meat, poultry	Headache Abdominal pain Diarrhea Nausea Fever Vomiting	Avoid cross contamination of foods Maintain correct storage temperatures Thorough cooking of food Good personal hygiene Proper sanitation practices
Staph (Staphylococcus aureus)	Human nasal passages or on human skin, transmitted by poor personal hygiene or uncovered sores	Nausea Cramps Vomiting Diarrhea	Good personal hygiene Time and temperature controls Do not sneeze or cough in food area Wash hands with sanitizers Cover sores or boils

Workplace practices and procedures

With so many bacteria around, what can you do to keep them from contaminating the food?

There are numerous practices already established by the foodservice industry and by your employer that help to minimize the possibility of contamination. When you look at them as a whole, you'll notice that they focus mainly on cleanliness and common sense.

While your employer will give you specific instructions for your particular job, this book will cover a number of safety measures you can use to control hazards. Separate chapters will discuss these subjects in detail:

- Cross contamination
- Personal hygiene
- Time and temperature controls
- Sanitation

Wrap-up

It's your job to provide good tasting, safe food to your customers. Because there are so many potential hazards that can contaminate food, every foodservice professional must constantly be aware of the hazards and be alert to preventing them. By following the safe food practices covered in this book and discussed by your employer, you can keep the food you serve hazard-free.

NOTES

Employee ____
Instructor ____
Date ____
Company____

REVIEW OF PREVENTING HAZARDS IN FOOD

1. Name three types of food hazards.
   a. Biological, nuclear, aquatic
   b. Biological, physical, chemical
   c. Biological, mental, emotional
   d. Physical, chemical, nuclear

2. A biological hazard can be:
   a. Bacteria
   b. Parasites
   c. Molds
   d. All of the above

3. Biological hazards are often called "microorganisms" because:
   a. They are so small that you need a microscope to see them.
   b. They are obvious to the naked eye.
   c. They are so small that you don't need to be concerned about them.
   d. None of the above.

4. Dangerous microorganisms in food affect this group most severely:
   a. Teenagers
   b. Adults aged 21 - 55
   c. Young children, senior citizens, and people who are ill
   d. Healthcare workers

5. Symptoms caused by contaminated food include:
   a. Headache, toothache, hair loss
   b. Nausea, vomiting, diarrhea
   c. Backache, insomnia
   d. Persistent itch, loss of appetite

6. Which one of these statements is true?
   a. As long as you wash your hands periodically, there is no risk of infecting the foods you touch.
   b. Time and temperature controls for food do not reduce the risk of bacterial growth.
   c. The microorganisms in food that cause illness are common and possibly deadly.
   d. Your own personal hygiene has no effect on the food you prepare or serve.

7. Food establishments use which of these controls to curb the growth of microorganisms?
   a. Proper sanitation practices
   b. Time and temperature controls
   c. Moisture controls
   d. All of the above

8. The temperature danger zone for food is 40° - 140°F?
   a. True
   b. False

9. Dangerous bacteria can either produce the toxin in the food before it is eaten or in a person's intestines after the food is eaten.
   a. True
   b. False

10. Common disease-causing microorganisms in food include:
    a. E. coli
    b. Salmonella
    c. Staph
    d. All of the above

HOW TO LIMIT CROSS CONTAMINATION

Just what is cross contamination? And why do you need to be concerned about it?

Let's start with an explanation of what "contamination" means. It is the act or process of soiling or infecting something through contact.

In the foodservice industry, contamination could be a chip of glass in a drink, chemical residue on a meat slicer, or *Salmonella* in chicken. These are examples of three different types of contamination hazards—physical, chemical, and biological.

Cross contamination is a major food safety concern. It can cause injury, illness, and even death. Cross contamination occurs when food is contaminated with something that doesn't belong in it. It's called cross contamination because it's contamination from one thing to another.

Your responsibility, as a foodservice professional, is to make sure you do everything in your power to prevent cross contamination.

And you can—by following established practices and procedures, by knowing what to look for, and by constantly being alert to possible hazards.

Physical hazards

Physical hazards are objects that can get into food during handling, but these objects do not belong there. They include chips of glass or dishware, pieces of packaging material, metal, and wood.

Eliminating physical hazards can be easy, if you know what to look for and use common sense. However, physical hazards come from a variety of sources, and they are not necessarily obvious.

As an example, imagine someone scooping ice cubes from a bin with a glass. Seems innocent enough, right? Well, this action is very hazardous! If the glass breaks even slightly, the broken glass chips can easily fall into the ice bin and go unnoticed. The chips could then be put into beverages for customers and cause serious injury.

Inspect broken or damaged containers, and call them to the attention of your supervisor

As another example, suppose the can opener you are using is worn. It can sprinkle metal shavings on the canned food you're about to use.

There are numerous ways physical hazards can be present. Here are some hazards to watch for:

- Wood from splintered pallets during shipping
- Broken or chipped metal from equipment or utensils, such as a broken knife blade
- Pieces of packaging material, such as cardboard or staples
- Jewelry (chains, earrings, rings, watches), or pieces broken off them
- Pens and pencils
- Pieces of broken dishes and glasses

As you can see, you always need to be on the lookout for physical hazards, especially since they can be easily prevented.

Chemical hazards

While physical hazards may be tiny, they are still visible. However, some cross contamination hazards cannot be seen in the food we serve. These can be chemical hazards or biological hazards. We'll discuss chemical hazards first.

Chemicals like cleaning solutions, sanitizers, and pesticides that do not belong in food are known as chemical hazards. Allergens (substances that cause allergies) are also considered chemical hazards; they include foods such as peanuts, eggs, fish, and milk.

Of course, foodservice facilities need to have cleaning supplies, detergents, and sanitizers to provide a clean workplace and to kill dangerous bacteria. However, these chemicals can be poisonous. To reduce the risk of cross contamination, store chemicals and cleaning supplies away from food.

Store chemicals and cleaning supplies away from food.

When it is necessary to clean and sanitize areas near where food is stored or served, take precautions not to let the chemicals come into contact with the food.

Pesticides are another serious food safety hazard. Only people trained to use them should handle them. If pesticides are kept in your facility, they must be stored away from food-contact surfaces and other chemicals, because many pesticides contain chemicals that are poisonous to people and pose a serious cross contamination threat.

Pesticides can already be on some food, such as fresh fruits and vegetables. Be sure to wash them thoroughly to remove surface pesticide residue and dirt.

Biological hazards

Biological hazards include viruses, mold, parasites, and bacteria. Some of the most common and dangerous are *E. coli*, Hepatitis A virus, *Listeria*, *Salmonella*, and *Staph*.

Unlike physical and chemical hazards, whose source you may be able to see, bacterial hazards are invisible. To further complicate things, they travel, or cross contaminate, through numerous ways: by water, food, insects, and animals—especially by humans.

Bacteria are so small, we cannot see them. But they are everywhere—on our skin, in our hair, under our fingernails, on our clothes, and on the money we handle. If they entered our food, we wouldn't even know it (not right away, at least). The food could look, smell, and taste great, yet still it could be contaminated with harmful bacteria.

What can we do to prevent bacteria from hurting us and our customers? Many things. First, be sure to learn and follow your employer's food safety rules and procedures. Second, keep in mind the guidelines in this book.

To minimize contamination of food:

- Keep raw food and cooked food separate.
- Wash your hands thoroughly before handling food.
- Do not lick your fingers, wipe or blow your nose, or scratch yourself while preparing or serving food.

- Wash your hands thoroughly after handling raw beef, pork, or poultry.
- Wash fruits and vegetables completely.
- Use utensils with long handles for serving food.

Learn and follow food safety procedures at your workplace.

To keep bacterial growth to a minimum:

- Keep foods out of the temperature danger zone (learn more about this subject in the chapter "Time and Temperature Controls"). Essentially, keep hot foods hot and cold foods cold.
- Thaw foods using proper methods.
- Do not cool leftovers at room temperature.

Cross contamination most often occurs when raw food, especially meat, comes into contact with a ready-to-eat food. To prevent cross contamination as much as possible:

- Take extra care when working with raw beef, pork, and poultry (they are most likely to contain dangerous bacteria).
- Wash your hands after handling these raw products.
- Clean and sanitize countertops, cutting boards, utensils, carts, trays, and any additional work surfaces that have come into contact with raw meats before placing any other food on the same surface.

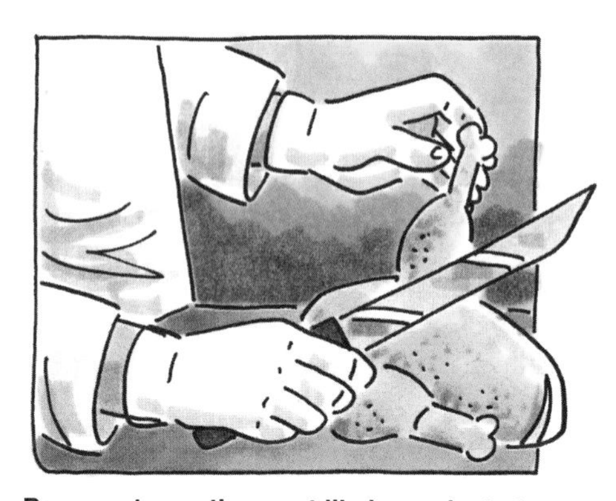

Raw meats are the most likely products to contain harmful bacteria. Take extra care when working with them.

- Do not re-use wash cloths or sponges after wiping counters with them. These items provide the ideal place for bacteria to grow. Wash the cloths and sponges instead. (Paper towels are the best choice for wiping up meat juices.)

Personal cleanliness

By keeping yourself clean and following strict cleanliness rules at work, you can keep the food you work with free from potential contamination. Your personal hygiene can make the difference between a safe food and one that causes illness or even death to those who eat it.

You need to take certain precautions (listed below) to eliminate the risk of causing illness. For more details on good personal hygiene habits, see the “Personal Hygiene” chapter.

Wash your hands

Washing your hands is a key way to prevent cross contamination. Wash your hands after any activity that could pose a cross contamination hazard, such as:

- Using the restroom
- Taking out the garbage
- Touching your skin
- Scratching your head
- Going from raw to cooked food
- Handling waste or spills
- Cleaning tables and equipment
- Touching money

Hands should be washed only in designated sinks and never in sinks used to prepare food. If you did so, the bacteria could remain in the sink and contaminate food. (See the proper procedure for handwashing in the "Personal Hygiene" chapter.)

Wash your hands when going from raw food to cooked food.

Wear gloves

Wearing gloves helps prevent cross contamination from bacteria on our skin. But wearing gloves is not a complete protector against it.

Gloves act as a "second skin" for hands, but gloves can easily become contaminated and spread contamination onto food contact surfaces or directly onto the food itself, and, ultimately, to your co-workers and customers.

For example, say you're wearing gloves and working the automatic dishwashers. You're putting away dirty dishes into one dishwasher, and then you must unload and stack clean dishes from another. To prevent cross contamination, you must not touch the clean dishes with the gloves you were using to load the dirty dishes. You need to put on a new pair of gloves.

When you do replace your gloves—after they've become ripped, torn, or soiled—throw the old pair away. Then wash your hands before putting on a fresh pair.

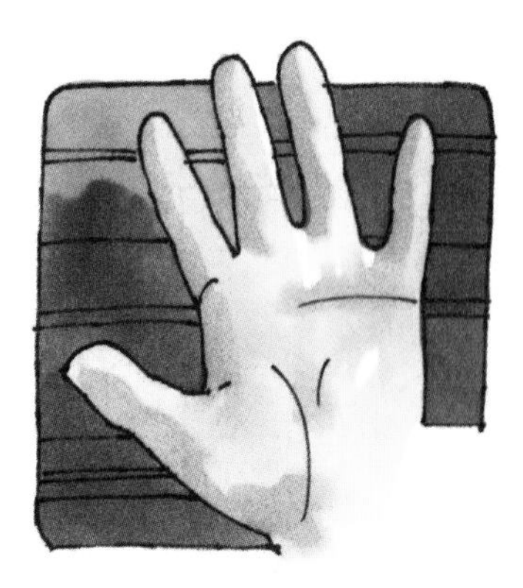

Gloves help to protect against cross contamination.

If you have a cut or sore on your hands, first cover it with a waterproof bandage. Then put on a fresh pair of gloves. The glove will keep the wound clean and away from the food. It will also protect the bandage from coming loose and falling into the food.

Use good personal hygiene habits

There are some additional ways you can directly help prevent cross contamination by maintaining good personal hygiene. Take daily showers before work, keep your fingernails trimmed and clean, wear your hair either short or pulled back, and wear a hat. In some places, hair and beard nets are also necessary.

Other things to remember are:

- Do not wear jewelry, false nails, or false eyelashes because they could fall into the food.
- If you are feeling sick or have an illness, do not come in to work. Think about the risk you pose to your coworkers and customers: you may pass the bacteria onto the food and spread your illness to them.

If you are ill, do not come in to work.

- Make sure your uniform and work clothes are clean. Not only do you want to avoid bringing contaminants into food preparation areas, but you also want to send the proper message to your customers—that you are a clean and professional person providing safe, quality food.

Additional points

No matter what type of cross contamination hazard you may encounter, there are things you can do every day to reduce the risk of injury or illness.

- As food is unloaded, check its temperature. Also check for large ice flakes or pools of water, which are signs of thawing. Be alert for odors and unusual smells. Any questionable deliveries should be immediately brought to the attention of your supervisor.
- Keep your receiving dock clean, and immediately transfer food on sanitary carts to the proper storage location.
- Always put food in the correct storage area and make sure labels are visible.
- Store raw meat, poultry, and seafood on the bottom shelf of the refrigerator to prevent any leaking juices from contaminating cooked food. Ready-to-eat foods must never be stored below raw foods.

Raw meat, poultry, and seafood belong on the bottom shelf of the refrigerator. This prevents their juices from spilling onto cooked food.

- Label all ready-to-eat foods with preparation dates and times, and serve this food on a “first in, first out” or “FIFO” basis. This means that the items that were prepared earliest need to be used first to reduce the chance of spoilage.

Label ready-to-eat foods with dates and times. Serve this food on a “FIFO”—first in, first out—basis.

- Always thaw frozen food in a refrigerator, microwave, or cold water that is changed frequently to keep it cold. Never thaw food at room temperature, where bacteria can grow quickly, and never refreeze thawed food because freezing does not kill bacteria; it only slows down its growth. If you have thawed food improperly and refreeze it, you still have a hazard.

- Keep work surfaces clean. After preparing any raw food, always clean and sanitize work surfaces, cutting boards, and knives before preparing any other food on the same surface or with the same utensil.

- Keep food at proper temperatures to slow bacterial growth. Food cartons should not be stacked too close together. Leave room for cold air to circulate and cool the food.

Wrap-up

Cross contamination is dangerous and poses a real health risk. There are numerous ways we can prevent it, however. Mainly you have to think about every step you take in the workplace. Keep in mind that what you do may affect the food you prepare or serve, and may ultimately harm others. Be sure to understand safe food practices completely, and follow them every day.

Employee ____________________
Instructor ____________________
Date ____________________
Company ____________________

REVIEW OF HOW TO LIMIT CROSS CONTAMINATION

1. "Contamination" means:
   a. The act of nurturing a living organism.
   b. The act of replacing a worn, used article with a new one.
   c. The act of infecting something with a harmful substance through contact.
   d. None of the above.

2. Which of these is an example of cross contamination?
   a. *Salmonella* in chicken is passed on to raw vegetables when the knife that first was used to cut the chicken is now used to cut vegetables.
   b. A foodservice employee sneezes into his hands and continues to knead dough.
   c. Neither a nor b.
   d. Both a and b.

3. Pieces of glass, chips of metal, and broken jewelry are all part of this group:
   a. Physical hazards
   b. Chemical hazards
   c. Biological hazards
   d. Nuclear hazards

4. Chemical hazards include cleaners, sanitizers, and pesticides.
   a. True
   b. False

5. Bacteria are most often present in:
   a. Raw vegetables
   b. Properly cooked food
   c. Raw beef, pork, and poultry
   d. Orange juice

6. Safe food practices call for:
   a. Keeping raw food and cooked food separate.
   b. Washing your hands after handling raw meats.
   c. Placing raw meats and raw seafood on the bottom shelf in the refrigerator.
   d. All of the above.

7. Wearing gloves is a very reliable way to prevent cross contamination.
   a. True
   b. False

8. Refreezing thawed food kills bacteria.
   a. True
   b. False

9. "FIFO" stands for:
   a. Food in front out first
   b. Freeze it, forward over
   c. First in, first out
   d. Fruit in, fat out

10. If you smell an unusual odor from delivered goods, you should:
    a. Immediately call it to the attention of your supervisor.
    b. Ignore it.
    c. Check the package yourself.
    d. Ask the driver when the food was first packaged.

PERSONAL HYGIENE

Personal hygiene can be a sensitive subject. However, it is so important to food safety, that every foodservice employee must be very sure to follow good personal hygiene habits.

Generally speaking, good personal hygiene involves keeping yourself and your clothes clean, following certain cleansing procedures at the appropriate times (such as handwashing), and following additional sanitary work practices when you work with and around food.

While a certain action—touching a pimple, for example—might be considered rude or offensive in another setting, this same action is actually dangerous when it takes place in a foodservice environment. If that foodservice employee does not wash his hands after touching a pimple, then he could spread bacteria into food and make people sick.

This chapter discusses the personal hygiene habits you need to follow every day to ensure your own safety, the safety of the food you handle, and the safety of the customers you serve.

Handwashing

If you were to identify the single, most critical part of personal cleanliness in foodservice, it would have to be frequent and thorough handwashing. That's because dirty hands and fingernails are the most common carriers of bacteria. You need to keep in mind that thorough handwashing must follow any action you do that might cause your hands to become contaminated.

When to wash your hands

Be sure to wash your hands at these times:

- Immediately before putting on gloves.
- After coughing or sneezing into your hands.

- After touching or scratching areas of the body, such as ears, mouth, nose, or hair.
- After you adjust your clothing, such as aprons, hats, hairnets, shoes.
- After using a tissue or handkerchief to wipe or blow your nose.
- After eating, drinking, or smoking.
- After using the restroom.

Always wash your hands after using the restroom.

- After touching unclean equipment or work surfaces.
- After picking up an item off the floor.
- After clearing away and scraping used dishes and utensils.
- Before and after handling raw meats, poultry, or other raw foods.
- After handling items such as garbage, brooms, hoses, soiled wash rags, etc.
- Before touching clean dishware, tableware, and flatware.

- After working with any type of allergenic food (like peanut butter or walnuts).

How to wash your hands

It seems pretty easy to wash your hands. After all, you've been doing it for years. But when you're a foodservice professional, proper handwashing requires more than simply soaping your hands under running water. It calls for these steps:

Turn the water on to a temperature as hot as your hands can tolerate, and wet your hands under the running water.

Soap up your hands with the antibacterial soap provided, lathering beyond the wrists (and up to the elbows if you are wearing short sleeves).

Scrub your hands vigorously against each other for 20 seconds, making sure that the soapsuds cover and clean every part of your hands. Pay particular attention to the webbing between your fingers and the area around your nails, where dirt and germs can hide. Make sure you clean under your fingernails, scraping any dirt out from under them.

Rinse your hands thoroughly under the hot running water. Let the water flow from the elbows to the fingertips.

Turn off the faucet with your elbow or a single-use paper towel, and dry your hands with the disposable, single-use paper towels provided.

Drying your hands on your apron or scratching your head will mean you'll have to wash again. In addition, don't use hand lotion on your freshly washed hands, because its moisture can cause bacteria on the skin to grow.

Serving food in a sanitary way

Suppose every safety precaution has been taken to prepare food. The food itself is safe. Still, there will be one additional step that could contaminate it—and that's how you serve the food. Contamination from employees or utensils may pose a danger. So there are certain methods of serving food properly, and a number of points to remember:

Right
Wrong
Right
Wrong
Right
Wrong
Right
Wrong
Right
Wrong
Right
Wrong
Right
Wrong
Right
Wrong

- If you handle dishes, glasses, and flatware, you must never touch the part that will come into contact with the customers mouth. Hold plates by the bottom or at the very edge, flatware by the handles, and cups or glasses by the bottoms or handles.
- Never touch food with your bare hands. Use a serving utensil, such as tongs to pick up breads and rolls.
- If you handle money, don't touch food after touching money until you have washed your hands. Coins and bills are handled by so many people, that they are full of bacteria.
- If you clear tables, be sure to wash your hands thoroughly before handling clean place settings or serving food.

Maintain other clean habits

While everyone should try to maintain personal cleanliness, it is absolutely essential for you, as a foodservice employee. Customers often judge a food establishment by observing the personnel serving them. Even more important, if you are not clean, you could contaminate the food.

Personal cleanliness begins at home. Follow these simple habits to maintain necessary personal cleanliness:

- Shower or bathe every day before you go to work. Be sure to keep hair clean as well, since dirty hair can carry bacteria.

Take a shower or bath every day before coming to work.

- After showering, look for any sores or cuts, especially on your arms and hands. If you find any, cover them with waterproof, plastic bandages. (Colored bandages are more noticeable, and serve as a reminder to watch for the bandage coming loose.) At work, wear a glove to cover the bandage.
- Put on clean, laundered clothes.
- Use soap and water to wash hands thoroughly every time you use the restroom, at home or on the job.
- Trim, file, and clean fingernails frequently.
- Remove nail polish, fake nails, and jewelry before starting work. These can break off and fall into food.

Stay home if you are sick

With all the emphasis on cleanliness and food safety in this book, you might think that everyone already knows the following rule, but it does need to be pointed out: If you are ill, do not come in to work.

The reasons are obvious. Not only will you slow down your

own recovery, but you will also spread your germs to others. Remember, never go to work if you are contagious with bacteria or a virus that could be transmitted to others through food or water.

Do not come in to work if you have a contagious illness.

If you have a sore throat, fever, cough, sinus pain, or other symptoms of a cold or the flu, recognize that these are signs of infection and pose a hazard in your workplace. Intestinal symptoms, such as diarrhea, abdominal cramps, and vomiting also can be dangerous. So do yourself, your co-workers, and your customers a favor—stay home if you have any of these symptoms.

Above all, follow your company's policy on when to call in sick.

Wear clean outer garments

Your company will tell you the kinds of clothes you must wear while working. You may even be provided with a uniform, apron, hat or cap, shoes, or other clothes for your job. Wear your clean uniform only at work to help prevent contamination.

Also, make sure your uniform is clean when you come to work. Not only will this help prevent contamination, it will also serve another purpose. It will reflect well upon you and your company. If you were to wear dirty clothes, customers would suspect that the food you serve might be unsanitary.

Keep aprons and other outer garments free from possible contamination by removing them before using the restroom and leaving them outside the restroom. That way you can go to the bathroom, wash your hands thoroughly, and put your outer clothing back on without contaminating it.

Gloves, hairnets, and hats

Some jobs require the use of gloves. If yours is one of them, keep in mind that wearing gloves does not substitute for handwashing. Always wash your hands thoroughly before putting on gloves.

Gloves also can spread bacteria in the same way that bare hands do. So change your gloves whenever you would wash your hands. In addition, if your gloves get dirty or torn, toss them out and put on a fresh pair.

When gloves get dirty or torn, throw them away and put on a fresh pair.

Foodservice employees are often required to wear hairnets. Your company will probably provide these, or ask you to get a particular type. In either case, you are responsible for wearing them effectively, making sure the hairnet covers your hair completely, so none can escape and fall into the food. If you have long hair, you'll find it easier to keep your hair contained by first putting it in a ponytail or braid. This way, strands are less likely to come loose.

If you have a beard, a restraint should be worn for the same reasons as a traditional hairnet. Special beard covers are available to restrain beard hair, and should be worn along with traditional hairnets or other head coverings.

Don't perform any personal grooming, like combing your hair, while working with or around food. Try not to touch your hair again until after your work shift is complete. If you do have to touch any hair, because of an itch or by accident, go to a restroom or changing area and:

- Make sure your hair is intact under its hair restraint.
- Wash your hands before returning to your work area.
- Change gloves if you are wearing them.

Where to store your belongings

Your company may provide an employee lounge, break area, or changing area where you must keep non-work clothing and other personal belongings. These areas need to be separate from areas where food is exposed or where equipment or utensils are washed. Belongings that should be kept in these designated areas include your purse or wallet, jewelry, street clothes, coats, cigarettes, books, magazines, and other non-food items.

Don't bring your lunch, soda, snacks, or other personal food items into these areas, because they could attract pests and may increase the risk of contaminating work areas. These personal food items belong in an area designated for that purpose.

Wrap-up

Customers trust you to provide them with food that is safe to eat. Do your part by following these personal hygiene habits:

- Shower or bathe daily.
- Keep your fingernails clean, trimmed, and filed.
- Cover cuts and sores with a bandage.
- Make sure work clothes are clean and freshly laundered.
- Wear a hairnet or hat, if required.
- Remove all jewelry.
- Wash your hands often and thoroughly.
- Stay home if you are ill.

NOTES

Employee ____
Instructor ____
Date ____
Company____

REVIEW OF PERSONAL HYGIENE

1. Hygiene mainly deals with:
   a. Safety
   b. Medicine
   c. Cleanliness and health
   d. Treating wounds

2. What is the single, most important act of personal cleanliness in foodservice?
   a. Wearing clean clothes
   b. Washing your hands thoroughly and frequently
   c. Removing jewelry
   d. Wearing gloves

3. When must you wash your hands?
   a. After blowing your nose
   b. After using the restroom
   c. After handling raw meat
   d. All of the above

4. What is the least amount of time to scrub your hands when you are washing them?
   a. 20 seconds
   b. 30 seconds
   c. One minute
   d. Two minutes

5. Which statement is correct?
   a. Carry glasses from the top.
   b. Hand a spoon to a customer by holding the bowl of the spoon while the handle points to the customer.
   c. Always wash your hands after handling money.
   d. It's fine to serve rolls with your bare hands.

6. What should you do before coming to work?
   a. Shower or bathe.
   b. Cover any cuts or sores with a bandage.
   c. Put on clean, laundered clothes.
   d. All of the above.

7. Don't come to work wearing nail polish, fake nails, or jewelry because these items may break off and fall into the food.
   a. True
   b. False

8. Hairnets are required so that hair does not fall into food.
   a. True
   b. False

9. Looking clean and professional as a foodservice employee matters to customers.
   a. True
   b. False

10. Your lunch, snacks, soda, and other personal food items should be stored:
    a. With your other personal belongings.
    b. With food prepared for customers.
    c. In an area designated for employee foods.
    d. All of the above.

TIME AND TEMPERATURE CONTROLS

In the foodservice profession, you handle lots of food every day—from receiving, to storing, to cooking, and finally to serving. You even have leftovers. How do you store different types of food? And how do you keep those foods safe?

That's what this chapter is about. We'll discuss the proper temperatures that keep food safe, and the best practices for storing certain types of foods.

Temperature danger zone

Controlling the temperature of food is a critical practice in keeping food safe. That's because all food has some bacteria that can grow rapidly within a certain temperature range. In fact, temperature control is the one factor most often mentioned in connection with people getting sick from food. This is unfortunate, because we have the know-how to prevent this sickness from happening.

Keep foods out of the temperature danger zone—40°F to 140°F.

Since disease-causing bacteria are known to multiply quickly at temperatures between 40°F and 140°F, this range is called the "temperature danger zone." No food should be allowed in the temperature danger zone for more than two hours. Keep in mind, this is a total time limit, including each stage of handling and serving. These times add up quickly!

It's also important to note that bacteria grow more rapidly in the middle of the danger zone —70°F to 120°F (think of this as room temperature).

As a foodservice employee, you need to make sure that you keep food out of the temperature danger zone, whether your job is to unload food shipments, store food, cook and display food, or serve food.

At the loading dock

Let's start at the loading dock. As soon as a shipment comes in, you need to check the internal temperatures of cartons of refrigerated and frozen food. Select cartons from various places in the truck and check the internal temperature of the food with a thermometer. (Make sure the thermometer is cleaned and sanitized before and after each use.)

Also, look for signs of thawing, which may mean the food was exposed to warm temperatures during shipment. If you notice large ice crystals or pools of solid ice, that might indicate that the ice had melted and then refroze. If you suspect a problem, tell your supervisor immediately.

The food that arrives will need to be stored. Most foodservice businesses have a dry storage area, refrigerators, and freezers. These also need to be checked on a regular basis.

Refrigerators

Refrigerators help keep food out of the temperature danger zone, and therefore slow bacterial growth. This increases the shelf life of most foods.

Generally speaking, the colder a food is, the safer it is. A maximum refrigerator air temperature of 40°F or lower must be consistently maintained to keep the food inside safe. Be sure to check the temperature of the refrigerators every day.

Check the refrigerator temperature daily. It needs to be 40°F or lower.

Other helpful hints:

- Since there are optimal storage temperatures for each major food category, the ideal situation would have a separate refrigerator for each group. Since that may not be possible, store meat, fish, and poultry in the coldest part of the refrigerator, away from the door.
- Try not to open the refrigerator door too often, and don't ever prop it open. Every time that door opens, cold air escapes and warm air enters, making the air warmer inside.

- Cold air needs to circulate around the food to keep food cold. So don't overload the refrigerator or stack food too close together.
- Store dairy products away from foods with strong odors.
- Store raw or uncooked food on the bottom shelf of the refrigerator. This prevents cross contamination from fluids dripping down.

Freezers

Storage freezers are designed to keep frozen foods at 0°F or below. Only food received in a frozen state should be kept in the freezer, unless the food is to be prepared that day.

Don't overload the freezer. Cold air needs to circulate around the food.

Many of the practices that apply to refrigerators also apply to freezers. Allow air to circulate by not overcrowding the freezer and by leaving room between foods. If foods are pressed too close together, the packaging can act as insulation and raise the temperature of the food.

Also, open the freezer door only when necessary and only for a short time to avoid raising the temperature inside. Check the thermometers in the freezer on a regular basis.

Dry storage areas

Areas for storing dry foods—such as rice, beans, and canned goods—also have certain requirements. The area should be clean, dry, protected from moisture and heat, and well ventilated. Good ventilation helps to preserve dry food.

Guidelines for all food storage areas

- Keep storage areas clean and dry to prevent contamination from dirt and pests.
- Use food in the same order it was received. This means that the food first in should be the first out. It's easy to remember as the FIFO system (first in, first out). This system ensures that older food is used before newer food. It also cuts down on spoilage and potential risk due to "expired" food.
- Because we use the FIFO system in foodservice, it's important to label food properly. Before storing any food, be sure to mark what the food is and the date and time it was received.

Label food with the date and time. Use the food on a first in, first out (FIFO) basis.

- Store food only in areas designed for that type of food, to eliminate risk of contamination and to keep foods within the correct temperature range.
- Do not put food near chemicals, heating vents, boilers, furnaces, or toilets. This helps to avoid cross contamination and changing the temperatures of food.

Thawing and cooling down

While freezing food keeps most bacteria from multiplying, the freezing process does not kill bacteria. In fact, once food is removed from the freezer, bacteria can grow very quickly if the food is allowed to thaw at room temperature. Be sure to thaw foods out of the temperature danger zone. Never thaw foods at room temperature.

These are the three safest ways to thaw foods:

- In the refrigerator.
- In the microwave.
- Under water at a temperature of 70°F or below for no more than two hours.

A safe way to thaw frozen food is in the refrigerator.

Leftovers require special attention. Hot foods need to be cooled down rapidly to minimize the growth of bacteria. The correct way to cool down food is to divide it into small batches and place it in shallow pans. Then place the pans in the refrigerator, allowing plenty of cold air circulation. Make sure you cover and label (product and date) each pan or tray of food.

When you are ready to use the leftover food again, reheat it rapidly and serve it immediately. Never mix leftover foods with fresh foods because the leftovers can transfer bacteria to the fresh food.

Serving and display areas

Serving and display areas call for specific guidelines. Keep these in mind:

- Prepare only as much food as you can serve in a specific period of time.
- Do not let food sit too long in coolers or hot serving containers (such as steam tables, double boilers, and chafing dishes). These foods may end up in the temperature danger zone.
- Keep food in hot holding equipment at a temperature of 140°F or higher. Check it periodically with a thermometer.
- Cover hot foods to keep them hot, and stir the food frequently to distribute the heat.

Wrap-up

One of the most effective controls you have in keeping food safe is to make sure food remains out of the temperature danger zone (40°F to 140°F). So you need to monitor the temperature and storage of food from the moment the delivery shipment arrives until the prepared food is served to your customers.

NOTES

Employee ________________________________
Instructor ________________________________
Date ________________________________
Company ________________________________

REVIEW OF TIME AND TEMPERATURE CONTROLS

1. Room temperature is within the temperature danger zone.
   a. True
   b. False

2. The temperature danger zone is:
   a. 0°F–40°F
   b. 20°F–100°F
   c. 40°F–140°F
   d. 60°F–160°F

3. The maximum amount of time that food can be left in the temperature danger zone is:
   a. One hour
   b. Two hours
   c. Three hours
   d. None of the above

4. A good practice for storing raw meat is to:
   a. Store on the bottom shelf in the refrigerator away from the door.
   b. Store in the door of the refrigerator.
   c. Store on the top rack of the refrigerator away from the door.
   d. None of the above.

5. Freezer temperature should stay at 0°F or below.
   a. True
   b. False

6. Which statement is incorrect regarding food serving and display areas?
   a. Prepare more than the amount of food you can serve during a specified period of time.
   b. Don't let food sit for too long in coolers or hot serving containers.
   c. Keep food in hot holding equipment at 140°F or higher.
   d. To keep hot foods hot and distribute the heat, stir the food often and cover it.

7. In foodservice, a good practice is to use the older food in storage before the newer food.
   a. True
   b. False

8. The best way to thaw frozen food is:
   a. In the microwave
   b. In the refrigerator
   c. Under cold water for less than two hours
   d. All of the above

9. To cool down hot foods quickly, you should:
   a. Refrigerate in large container immediately.
   b. Separate into small batches and put in shallow pans before refrigerating.
   c. Freeze for one minute before refrigerating.
   d. Allow to cool at room temperature before refrigerating.

10. Mixing leftover food with fresh food can transfer bacteria to the fresh food.
    a. True
    b. False

SANITATION

The term "sanitation" may bring to mind thorough cleaning of equipment and work areas in your foodservice facility, but it goes beyond that. Sanitation refers to all the practices and step-by-step procedures necessary to keep your facility clean, and the food you serve free of contaminants and harmful microorganisms.

Good sanitation practices include:

- Personal hygiene and cleanliness,
- Cleaning and sanitizing equipment and work areas,
- Using and storing cleaning chemicals and supplies, and
- Pest control.

Personal cleanliness

One of your most important responsibilities in preparing and serving safe food is to be sure you come to work clean. That means paying attention to good personal hygiene every day. In fact, personal hygiene is so important that a separate chapter in this book is devoted solely to that subject.

In brief, good personal hygiene in foodservice requires you to bathe daily before work, cover cuts or sores, wear freshly laundered work clothes, and wash your hands often throughout your workday. In other words, keep yourself clean and make sure you don't contaminate any food you may serve to customers.

Wash your hands often throughout the day.

For more information on personal cleanliness, read the chapter called "Personal Hygiene."

Equipment and work areas

Personal hygiene is the first step in good sanitation practices. Beyond that, you need to keep the things you work with clean and sanitary. Food processing equipment, food contact surfaces, and food preparation utensils must all be cleaned and sanitized.

Think of it as a one-two punch against food contamination. First, you clean equipment and surfaces to remove soil and food residue. Second, you sanitize with a chemical that kills microorganisms. These two actions help to prevent bacteria and other harmful microorganisms that may be present on the equipment from contaminating the food.

Any food residue left on work surfaces, equipment, or utensils is a source for microorganisms to grow. Because they grow so rapidly, these microorganisms can contaminate the next food item that comes into contact with it. That's why it's important to remove all food residue before sanitizing. This way the sanitizer can do its job properly.

First clean, then sanitize equipment and surfaces.

If you have to clean and sanitize equipment, you probably have been trained to do so in a step-by-step process. It is critical that you follow the steps in the proper order, allow the cleaning and sanitizing agents to work for the specified period of time, and complete these steps as often as necessary.

Keep in mind that cleaning and sanitizing involve three basic steps:

- A rough cleaning, such as wiping off all visible residue with a cloth or brush.
- A thorough cleaning, with a cleaning chemical, like a detergent.
- Sanitizing with a chemical, like a bleach solution.

You may or may not have to rinse after the second and third steps, depending on which chemicals are used and at what strength. The chemical strength and contact time (the amount of time you have the cleaner or sanitizer in contact with the surface being cleaned or sanitized) are very important. Be sure to follow the directions listed on the chemical's label.

Whatever job you have in foodservice, you're probably involved in some way with cleaning and sanitizing. Be sure to follow the steps you've been trained to do. Perform these activities throughout the day to maintain sanitary conditions.

Cleaning chemicals and supplies

As necessary as cleaning supplies and chemicals are for sanitizing the food workplace, they can be a contamination threat. They must be used, handled, and stored properly. For example, bleach is a powerful sanitizer, but it can also create a toxic contamination hazard if left out or if stored too close to where food preparation takes place. If bleach accidentally gets into food, it would cause a serious health hazard.

Once you've cleaned the cleaning equipment, hang it up and let it dry completely.

Many other cleaning and sanitizing chemicals are also toxic and dangerous to humans if accidentally consumed. The cleaning tools and chemicals used for sanitizing must always be used according to label directions and your company's procedures. They must also be stored safely—away from food contact surfaces and equipment—in their designated storage areas, because, if not, the chemicals could end up in food.

What you actually need to do is clean the cleaning equipment before putting it away. Dirt or food left on mops or brushes will provide a source of food for microorganisms to grow. Then the next time you use that same mop or brush you will actually spread the microorganisms onto equipment and work surfaces, rather than removing it.

Once you have cleaned the cleaning utensils—including mops, brushes, and cloths—be sure to rinse and wring them out. Then hang them to dry. Allow these items to dry completely to prevent microorganisms from thriving in the damp and wet conditions. (Some companies use disposable cloths to avoid this problem.)

Pest control

Pest control is a common sense sanitation issue. No one wants pests—such as birds, rodents, cockroaches, or flies—near food because of their dangerous food contamination risk.

To keep food work areas safe and free of pests, follow your company's rules and these common sense rules:

- Keep personal food out of food work areas and clothing storage areas. Store your lunch, snacks, chewing gum, and candy in the designated areas so they don't attract pests.
- Never prop doors or windows open.
- Use pesticides only according to established procedures, and store them separately with cleaning chemicals.

- Clean up any spilled food or ingredients that can be an attraction to pests.

Store your lunch in the place designated for that purpose.

Wrap-up

Keeping the workplace clean and sanitary is every employee's responsibility. Start by practicing good personal hygiene, and extend the idea of clean and sanitary to the equipment you work with. Follow your company's procedures in a step-by-step manner to make sure you prepare and serve food that is safe to eat.

Employee ______________________________
Instructor ______________________________
Date ______________________________
Company ______________________________

REVIEW OF SANITATION

1. Good personal hygiene plays an important part in sanitation practices.
    a. True
    b. False

2. Sanitation in foodservice refers to:
    a. Cleaning of equipment and work areas
    b. Sanitizing of equipment and work areas
    c. Pest control
    d. All the practices needed to keep the workplace clean and the food uncontaminated

3. What is the proper order for cleaning and sanitizing?
    a. Rinse, sanitize
    b. Rough clean, sanitize
    c. Rough clean, clean with cleaner, sanitize
    d. None of the above

4. No pieces of food should be left on equipment because:
    a. It looks bad.
    b. It smells bad.
    c. It attracts pests.
    d. It is a source of food for microorganisms to grow.

5. Storing cleaning and sanitizing chemicals and supplies near food areas could cause a contamination hazard.
    a. True
    b. False

6. If you use a brush that has not dried completely after cleaning:
   a. It's perfectly fine.
   b. It could spread microorganisms that thrive in damp and wet conditions.
   c. It will ruin the brush.
   d. None of the above.
7. Cleaning tools need to be:
   a. Cleaned
   b. Cleaned and rinsed
   c. Cleaned, rinsed, and wrung out
   d. Cleaned, rinsed, wrung out, and hung to dry completely
8. When storing cleaning and sanitizing tools, you should:
   a. Clean the items thoroughly before storage.
   b. Dry the items completely before putting them away.
   c. Store the items only in their designated storage area.
   d. All of the above.
9. Pests that can contaminate food include:
   a. Birds
   b. Rodents
   c. Insects
   d. All of the above
10. Which of these rules applies to pest control?
   a. Never prop doors or windows open.
   b. Keep your lunch and snack items in the area designated for employee food.
   c. Both a and b.
   d. Neither a nor b.

HACCP

Have you ever heard of the term "HACCP" (pronounced há-sip)? It may sound like it stands for something foreign and formal, but the reason it exists is fairly simple. HACCP is a valuable system for food safety that focuses on preventing hazards in food.

It started in outer space

HACCP stands for Hazard Analysis and Critical Control Points (more on this later). The program was pioneered in the 1960s in connection with the U.S. space program.

The goal of the HACCP system then was to create food for U.S. astronauts that was virtually 100 percent risk-free. The government wanted to ensure against contamination that could cause illness or injury to its astronauts while they were in outer space. HACCP represented a new way of producing safe food.

HACCP was launched as a way to create safe food for U.S. astronauts.

Before the HACCP system was introduced, food was traditionally tested by sampling the finished product to identify high levels of contamination. HACCP aimed to identify and correct possible risks before they happened. The HACCP system would ensure that no defects were present in the food eaten by our astronauts by monitoring the food preparation process—from start to finish.

Today, HACCP is a recognized, proven system that has been adapted for use in foodservice operations nationwide. Your company may have its own version of the program, and you may already be aware of some of its parts. This chapter will introduce you to the overall system and explain the seven principles involved in applying the program to the place where you work.

The flow of food

Food travels a certain path in each foodservice facility. This is termed the "flow of food." It begins when your company receives food, and continues with storage, preparation, holding or display, and serving customers. But it doesn't stop there. You still have cooling, storing leftovers, and reheating as additional steps in the flow.

HACCP examines every step in the "flow of food," starting with receiving food products.

At each stage in the flow of food there are possible hazards that could cause contamination. The HACCP system works first toward identifying these hazards, and second toward coming up with ways to eliminate or reduce the hazards. It is a system of <u>preventive</u> controls that protects food, ensuring that food is safe to eat. A HACCP program focuses on continuous problem solving and prevention.

How HACCP works

Before getting into the specifics of a HACCP program, you need to understand that it is not a stand-alone system. It builds on other practices your company already uses to keep food safe. Good personal hygiene, proper sanitation, and pest control are all necessary to produce food that is safe to eat. In fact, these basic programs are important to have in place before setting up a HACCP system. Logic tells you that you have to be clean, use clean equipment, and work in a clean environment to produce safe food.

The "HA" in HACCP stands for hazard analysis. This part of the program examines every area where food is stored and handled, and looks for all the ways food can become unsafe to eat.

The "CCP" in HACCP stands for critical control points. These are the areas identified in the hazard analysis where food is at the greatest risk of becoming unsafe.

The seven principles

Seven principles guide the HACCP approach to food safety. Remember, the goal of these principles is to identify potential hazards and eliminate them from the process.

Principle 1: Conduct a hazard analysis

This first principle lays the foundation for the entire HACCP program. It is about understanding your complete foodservice operation, from beginning to end. This requires a flowchart (a step-by-step diagram) of the flow of food through all food handling and preparation processes.

Each step in the flow is then analyzed to determine where significant biological, chemical, and physical hazards might occur.

- Biological hazards include viruses, molds, parasites, and harmful bacteria.
- Chemical hazards could be cleaners, sanitizers, pesticides, and paint.
- Physical hazards include pieces of packaging material, chips from glasses or dishes, and bits of metal.

Principle 2: Identify critical control points (CCPs)

Through hazard analysis, CCPs are identified. A CCP, or critical control point, is a step or procedure in the food process at which a food safety hazard can be controlled. This means that if food safety precautions failed, the food could make someone sick.

CCPs may include cooking and cooling foods to specific temperatures, as well as storing foods correctly.

Principle 3: Set critical limits for each CCP

Each CCP must have boundaries that define safety. These are called their critical limits, and they are the maximum and/or minimum values of safety for each CCP. Critical limits set the standards for CCPs that provide you with the information to know when a food safety risk is present. A critical limit can be a cold storage temperature to reduce bacterial growth, or a cooking temperature to kill harmful bacteria.

Critical limits can be based on a variety and combination of factors, such as time, temperature, humidity, and the pH of a product. If food is outside its critical limit, it is known as a deviation, and the food is considered potentially unsafe.

Principle 4: Establish procedures to monitor each CCP

Once CCPs and their critical limits have been determined, someone needs to keep track of each CCP. That calls for monitoring, which involves making observations or taking measurements to see if the CCP is under control and is usually done following a specific procedure on a set schedule. For example, you might be responsible for checking the temperatures of the refrigeration and freezer units.

Each CCP needs to be monitored, such as the temperature in a refrigerator.

Principle 5: Establish corrective actions

This principle establishes the action you must take to correct a deviation in a CCP (when a critical limit is not being met). Essentially it addresses the question: What will you do when something goes wrong?

Your company needs to plan ahead and decide what needs to be done. This step is at the very core of HACCP. You can expect problems to arise. But if you notice them through monitoring, you'll be able to correct them before the problems cause a food safety hazard.

Principle 6: Establish a recordkeeping system

Records must be kept to document the HACCP system and to identify the areas that need improvement. These records can be simple, but they should include information on the hazard analysis, the CCPs, the results of monitoring activities, and how deviations for each CCP are handled. The records are especially valuable if a food safety incident occurs, because the records can help identify what happened and how large the problem may be.

An effective HACCP system depends on good recordkeeping.

Principle 7: Verify that HACCP is working

Suppose a HACCP system has been implemented—you've completed the hazard analysis, identified CCPs, set critical limits, established monitoring operations, determined corrective actions, and set up a record keeping system. The only thing left to do is to make sure it all works right. This is called verification.

Verification is a process that determines if the HACCP system is working as expected. It also looks for ways to improve the system. For example, in reviewing records, it is possible to notice trends in deviations that make you think about changing some of the steps in your workplace. Other verification procedures can involve the calibration of monitoring instruments, such as thermometers.

Wrap-up

At first glance, the HACCP system may appear complicated, but mainly it is a logical, effective, common sense approach to keeping food safe. The success of the program—and of your foodservice operation—depends on you. You need to recognize that you are part of a team whose goal is to make good tasting food that is safe to eat. HACCP will help you do this.

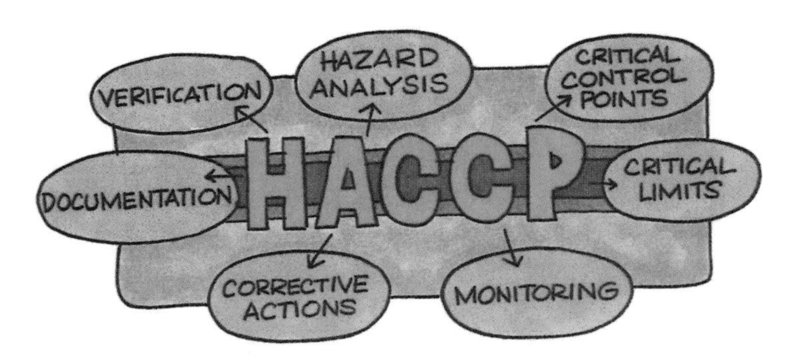

HACCP involves seven principles.

NOTES

Employee ______________________________
Instructor ______________________________
Date ______________________________
Company ______________________________

REVIEW OF HACCP

1. HACCP was first developed to ensure safe food for U.S. astronauts in outer space.
   a. True
   b. False
2. HACCP focuses on:
   a. Fixing problems after an illness arises.
   b. Random checks during the flow of food.
   c. Ongoing checks and problem solving to prevent unsafe food.
   d. Paper work.
3. "HACCP" stands for:
   a. Hazards are commonplace
   b. Hazard analysis and critical control points
   c. Hazards and allergens can cause problems
   d. Handle all critical control points
4. HACCP's approach to food safety involves how many principles?
   a. Three
   b. Five
   c. Six
   d. Seven
5. To conduct an effective hazard analysis you need to examine only the areas where food is cooked.
   a. True
   b. False

6. A critical control point is:
   a. A point in the flow of food where it is critical that certain actions are controlled.
   b. Each step in the flow of food.
   c. Both a and b.
   d. Neither a nor b.

7. Which is not a true statement?
   a. HACCP is the only system needed in foodservice to produce safe food.
   b. HACCP is a proven, effective system for keeping food safe.
   c. Critical limits must be set for each CCP.
   d. Recordkeeping is part of the HACCP system.

8. Keeping a freezer's air temperature no warmer than 0°F is an example of a:
   a. Physical hazard
   b. Chemical hazard
   c. Biological hazard
   d. Critical limit

9. Which of the following is a basic principle of HACCP?
   a. Conduct a hazard analysis.
   b. Set critical limits for each CCP.
   c. Establish a recordkeeping system.
   d. All of the above.

10. If food is outside its critical limit, it is known to be:
    a. A hazard
    b. Unsafe to eat
    c. A deviation
    d. Safe to eat

CUTS AND KNIFE HANDLING

A kitchen can be a dangerous place—whether it's at home or at a foodservice facility. In a busy kitchen, with professional food handlers working very fast, you really have to take care and follow safe practices. This is particularly true when you handle sharp knives.

Knife handling

It is often said that a cook's best friend and most important tool is a good sharp knife. While knives may be very valuable in the cooking process, they also pose certain dangers. Here are some helpful hints to keep in mind when you are working with knives.

Start with a good knife. A good knife should be well balanced and heavy, not easily bendable. If cared for properly, it will last a long time.

Keep your knives sharp. Although this may be hard to believe, a sharp knife is actually a safer knife. That's because it gives you control. It will cut through food easily, with little force. A dull blade, on the other hand, requires you to use force, which can cause loss of control and send your knife in an unpredictable direction.

Use a cutting board that gives you plenty of space for cutting.

Use a cutting board. Surfaces other than cutting boards can cause slipping and sliding. It's a good idea to put a damp towel under the cutting board to keep it from slipping while you work.

Choose the proper cutting board. Plastic cutting boards usually dull knives. Wooden boards will not. However, wooden boards have pores and nicks that trap bacteria, and so require thorough cleansing to remain sanitary. Whichever board you choose, make it large enough so you have plenty of space for cutting.

Use your knife correctly. Keep in mind that the tip of the knife—the most delicate part—is the section to use for cutting tender food, such as mushrooms. The middle portion of the knife is the area most used, and does the majority of the work. The back part of the knife, or the heel, is for the heavy work.

Cut with the edge pointing away from you. This seems fairly basic, but it's worth remembering. When you use a knife, make sure the edge points away from you and your fingers. This way, if you slip a bit, the blade will not keep going toward you and cause injury. Pay attention to where the knife's edge is pointing.

The edge of the knife should always point away from you.

Never cut something while holding it in your hand. Use a cutting board instead of "palming" the food.

Do not try to catch a falling knife. If you should drop your knife while working with food, stand back and let it fall. Fight the instinct to catch it. That could really hurt.

Use knives for their intended purposes only. Don't attempt to cut metal, paper, or string with your knives, and don't use them to open cans or bottles. Not only is it dangerous in terms of possible injury, but also in terms of cross contamination and ruining the knives.

Store knives in a knife holder. This will make it easier to grasp the handle of the knife you want to use. Also, it will keep the blades of the knives sharp. If knives are stored in a drawer, their blades can become damaged from banging around. Plus someone reaching into the drawer for a knife is more likely to get severely cut.

Take care when you clean knives. Wash and dry knives separately by hand. This will prevent rusting and dulling of the blades. More important, however, this practice will protect you. If knives soak in dishwater with other items, and you reach into the water, odds are you'll get a bad cut.

Wash and dry knives separately and by hand.

Watch where the edge of the knife points when you lay it down. The edge can be hard to see if it faces up, and someone may accidentally put a hand on it. When you lay a knife down, be sure the edge faces down.

Focus on what you are doing. Remember knives can be dangerous. Whenever you work with a knife, concentrate on the job at hand and practice knife safety habits.

Precautions for bloodborne pathogens

Another important issue regarding your safety when handling knives is the danger of exposure to bloodborne pathogens.

"Bloodborne pathogens" are defined as microorganisms present in human blood that can cause disease in humans. ("Pathogens" are microorganisms that cause disease.) Bloodborne pathogens include, but are not limited to, Hepatitis B and C, and HIV. These types of pathogens can be passed from one person to another through contact with blood.

If an accidental cut takes place and blood is present, take these precautions to prevent infection from bloodborne pathogens:

- Always wash your hands immediately after removing gloves and after any hand contact with blood. If a sink is not available, use an antiseptic cleanser, and then wash with soap and water as soon as possible.
- Clean and decontaminate equipment, utensils (including knives), and work areas as soon as possible after contact with any blood. Special disinfectants or cleaning solutions may be required.

- Never use your fingers to pick up broken glass or other sharp objects that have fallen. Use tongs or a broom and dust pan to clean up these items.

Don't use your fingers to pick up broken glass or other sharp objects. Use tongs or a broom and dust pan.

- Use "standard (or universal) precautions." That means you should assume all human blood is infectious. This is a good practice because many people who carry bloodborne infections have no symptoms and may not be aware that they have an infectious disease.

- Avoid spraying or splashing blood.

Personal protective equipment

Employers must provide, and you must use, personal protective equipment (PPE) when the possibility exists for exposure to blood. This equipment must not allow blood to pass through it to your clothes, skin, eyes, or mouth.

PPE must be accessible and available in appropriate sizes. It also must be kept clean and in good repair.

Single-use gloves must be replaced as soon as possible after they are contaminated or if they become torn or punctured. These gloves should never be reused.

Handling contaminated laundry

Once laundry has been contaminated with blood, it should be handled as little as possible. Laundry must be bagged where it was contaminated. Wet laundry must be placed in leak-proof bags.

All employees who handle contaminated laundry must wear gloves.

Wrap-up

Knives are valuable tools in the kitchen, but they also call for extra care. Be sure to keep knife safety practices in mind every time you handle a knife. Remember to follow the necessary precautions if you are exposed to human blood.

Employee ______________________________________
Instructor ______________________________________
Date ______________________________________
Company ______________________________________

REVIEW OF CUTS AND KNIFE HANDLING

1. A sharp knife is safer than a dull one.
   a. True
   b. False

2. Which of these statements is false?
   a. Using a cutting board as a cutting surface decreases the chance of slipping and sliding.
   b. Wooden cutting boards are dishwasher safe.
   c. Plastic cutting boards can dull knives.
   d. Keeping knives sharp is a good safety practice.

3. The back part, or the heel, of the knife is best suited for:
   a. Delicate cutting.
   b. The majority of the cutting work.
   c. The heavy work, such as cutting the thick stems of broccoli.
   d. Opening a bottle.

4. As you use a knife, make sure the edge points:
   a. Toward your hand.
   b. Away from you.
   c. Toward the wall.
   d. North.

5. When you use a knife for cutting fruit, you should hold the fruit in one hand and cut into it with the other.
   a. True
   b. False

6. When you lose your grip on a knife, you should:
    a. Block it with your hand.
    b. Try to catch it.
    c. Scream.
    d. Stand back and let it fall.

7. Why should you wash and dry knives separately and by hand?
    a. To keep knives from rusting and becoming dull.
    b. To protect the person who is washing dishes from accidentally reaching into the dishwater and coming out with a cut.
    c. To focus on being extra careful as you handle knives.
    d. All of the above.

8. Knives should be stored in a knife holder rather than a drawer.
    a. True
    b. False

9. Knife safety is a matter of common sense. Mainly, you need to:
    a. Have plenty of space to work in.
    b. Use good knives.
    c. Pay close attention to what you are doing whenever you deal with knives.
    d. Use sharp knives.

10. Bloodborne pathogens include HIV and Hepatitis B and C.
    a. True
    b. False

SLIPS, TRIPS, AND FALLS

The foodservice business has a lot of activity. It seems people are moving quickly all the time—in the kitchen and in the dining area. With all of this movement taking place, you need to be careful not to slip, trip, or fall.

Your foodservice operation can get very busy and active. Take care not to slip, trip, or fall.

The slip, trip, and fall factors

You might think that an accident due to a loss of balance is fairly uncomplicated. But, actually, slips, trips, and falls involve three laws of science—friction, momentum, and gravity. Two of them, friction and momentum, you can do something about.

Friction is the resistance between things, such as between your shoes and the surface you walk on. Without friction, you are likely to slip and fall. For instance, a slip on ice, where your shoes can't "grip" the surface, is due to losing traction.

Momentum is affected by speed and size of the moving object. You've heard the expression, "The bigger they are, the harder they fall." You can interpret that to mean the more you weigh and the faster you are moving, the harder your fall will be if you should trip or slip.

Gravity is the force that pulls you to the ground once a fall is in progress. If you lose your balance and begin to fall, you are going to hit the ground. Your body has automatic systems for keeping its balance. Your eyes, ears, and muscles all work to keep your body close to its natural center of balance. A fall is likely to occur if your center of balance (also called center of gravity) shifts too far and cannot be restored to normal.

About slipping

Everyone has slipped at one time or another. Your feet slide out from under you, and you land on the ground. What causes this sudden loss of balance? Too little friction between your feet and the surface you walk on.

Slips can be caused by wet surfaces, spills, or weather hazards like ice and snow. Slips are more likely to occur when you hurry or run, wear the wrong kind of shoes for what you are doing, or don't pay attention to where you are walking.

You're more likely to slip on a slippery surface when you hurry.

How to avoid slips

Follow these safety precautions to avoid slipping.

- **Practice safe walking skills.** Sounds elementary, but be alert to slippery surfaces and the speed at which you walk. If you must walk on or over wet surfaces, take short steps to maintain your center of balance and point your feet slightly outward. Move slowly and pay attention to the surface you are walking on.
- **Clean up spills right away.** Even minor spills can cause a slip. Whenever you see a spill (such as food or beverage), clean it up yourself or ask the person who is responsible for cleanup to do it. However, if you know the spill is hazardous or you are unsure what has been spilled, tell your supervisor immediately.
- **Always pick up the items you drop or that someone else has dropped.** Food items can be especially slippery.
- **Be more cautious on smooth surfaces.** Move slowly on floors that have been waxed but not buffed, and other very slippery surfaces.
- **Wear the right shoes.** The best way to prevent slips is to increase friction between your shoes and the surfaces you walk on, and increase the amount of surface of your shoe that touches the floor.

Wearing shoes with non-skid soles will decrease your chance of slipping.

- **Wear boots in snow, ice, and rain to avoid slipping in the parking lot.** Change into your work shoes once you are safely inside the building.
- **Use rubber mats as permanent or temporary solutions to slippery areas.** Use them in places known to be slippery, such as near the sinks and food preparation areas.

Use rubber mats in areas that are known to be slippery.

About tripping

Trips occur whenever your foot hits an object and you are moving with enough momentum to be thrown off balance. A trip can happen when your work area is cluttered, when lighting is poor, or when an area has loose or uneven footing. Trips are most likely to occur when you are in a hurry and don't pay attention to where you are going. Remember these rules to avoid tripping:

- **Make sure you can see where you are going.** If you are carrying a load, be certain you can see over the top of it.

When you carry something, be sure you can see over the top.

- **Keep work areas well lit.** Turned off lights and burned out bulbs can interfere with your ability to see clearly.
- **Keep your work area clean.** Don't clutter pathways, aisles, or stairs. Store all foods and other materials in their designated areas promptly.
- **Report to your supervisor any hazards you notice.** They can include loose carpeting, broken treads, or loose floor tiles.

About falls

Falls happen whenever you move too far off your center of balance. Slips and trips often push you off your center of balance far enough to cause a fall, but there are other reasons you may fall.

Makeshift ladders can cause falls, such as climbing on a chair or counter to reach something. Misuse of real ladders can also cause a fall. Most falls are slips or trips at ground level, but falls from greater heights can result in serious injuries.

Don't stand on make-shift ladders to reach something. Falls from high up can cause serious injuries.

Wrap-up

Slips, trips, and falls can be preventable if you are aware of your surroundings and use common sense. Essentially, this means wear the proper shoes, move slowly on slippery surfaces, keep your work areas and aisles clutter-free, clean up any spills, and pay attention to where you're going.

Employee ______________________________
Instructor ______________________________
Date ______________________________
Company______________________________

REVIEW OF SLIPS, TRIPS, AND FALLS

1. Of the three laws of science that affect slips, trips, and falls, which can you control?
   a. All three—friction, momentum, and gravity.
   b. Friction and momentum.
   c. Momentum and gravity.
   d. None of them are controllable.

2. What is a good example of using friction to avoid falling?
   a. Use rubber mats in slippery areas.
   b. Wear non-skid shoes in the workplace.
   c. Both a and b.
   d. Neither a nor b.

3. Momentum affects a fall with:
   a. Size and speed
   b. Size only
   c. Speed only
   d. Posture

4. What can you do about momentum to keep from falling?
   a. When you carry a heavy load, walk slowly.
   b. When you must walk on a slippery surface, walk slowly with small steps.
   c. Both a and b.
   d. Neither a nor b.

5. Avoid slips and trips by:
   a. Picking up any items that have been dropped on the floor.
   b. Walking slowly when you cannot see over the load you are carrying.
   c. Feeling your way around in the dark.
   d. None of the above.

6. During a slip or trip, the human body automatically tries to maintain its:
   a. Posture
   b. Momentum
   c. Composure
   d. Center of balance

7. If you spill something during a busy time, you should:
   a. Finish what you're doing and clean up the spill later.
   b. Clean up the spill right away.
   c. Ignore the spill until the maintenance crew cleans it up.
   d. Tell your co-workers to be careful.

8. When you are unable to reach something high up, it's a good idea to stand on a makeshift ladder, such as on a chair.
   a. True
   b. False

9. Most falls happen at ground level. But falls from higher levels can cause serious injuries.
   a. True
   b. False

10. Trips are most likely to happen when you're in a hurry, and when:
    a. Your work area is cluttered.
    b. The floor has loose tiles.
    c. The area is poorly lit.
    d. All of the above.

LIFTING TECHNIQUES TO SAVE YOUR BACK

What do you suppose is the largest single cause of back pain and injury? If you guessed car accidents, or sports, or falling, or even stress, you'd be offering some good possibilities. However, the fact is that improperly lifting moderate to heavy objects is the most common cause of back pain.

Fortunately you can do something to avoid back pain from improper lifting. This chapter discusses the correct lifting techniques you should use and other ways to save your back.

Let's start with some basic things you can do to keep your back strong and to prevent back pain.

Practice good posture

Whether you're standing, sitting, or lying down, your posture affects the amount of strain you put on your back. Poor posture increases strain on the back muscles and may bend the spine into positions that will cause trouble.

Sleeping on your stomach can cause a backache in the morning.

When you stand correctly, the spine has a natural "S" curve. Your shoulders are back and the "S" is directly over your pelvis. Your back should not be arched.

Good sitting posture has your knees slightly higher than your hips. Your hips should be to the rear of the chair, with your lower back not overly arched. Also, your shoulders and upper back are not rounded.

Reclining posture is important too. Sleep on your side with knees bent, or sleep on your back. If you sleep on your stomach, especially on a sagging mattress with your head on a thick pillow, you'll put too much strain on your spine, and you'll wake up with a backache.

Work on your physical condition

Your physical condition also can affect your back. If you are overweight, and particularly if you have a potbelly, you are putting extra strain on your spine. One way to look at it is that every extra pound up front puts 10 pounds of strain on your back.

Every extra pound in the front part of your torso puts 10 pounds of strain on your back.

Infrequent exercise can be a major factor in causing back pain, too. A sudden strain on generally unused back muscles leads to trouble, especially when you suddenly twist or turn your back. Proper diet and exercise is the sensible way to help avoid back problems.

Stress is another factor that may play a role. Tied in with your physical condition, stress can cause muscle spasms that affect the spinal nerve network. Although stress is a part of everyone's life, and a certain amount of stress is normal, excessive stress causes backache. So try to maintain a balanced lifestyle, and allow time to relax.

Avoid repeated strains

People often think back injuries result from lifting a heavy or awkward object. However, many back injuries do not come from a single lift. They are the result of relatively minor strains repeated over time. If you repeat a particularly irritating movement, the minor strains begin to accumulate and weaken affected muscles or ligaments. Eventually a more serious injury may result.

If you stand most of the time, place one foot on a ledge and switch foot positions often.

Ease the strain

Your job may require long hours of standing or sitting. These conditions can create back troubles. Be sure to get up and stretch frequently if you have to sit for long periods of time. If most of your work calls for standing, ease the strain on your lower back by placing one foot on a rail or ledge and switching foot positions often. However, keep your weight evenly balanced when standing. Do not lean to one side.

Basics of good lifting

There are numerous machines and equipment today to help you lift heavy objects. Forklifts, hoists, and dollies can make a lot of jobs easier. Often, though, especially in foodservice, it is necessary to load and unload moderate to heavy objects by hand. When this is the case, follow the proper steps for lifting (listed below), and you'll save yourself a great deal of pain and misery from an injured back.

1. **Size up the load before you try to lift it.** Test the weight of the load by lifting at one of the corners. If the load is too heavy, or if it has an awkward shape, you can:
    - Get help from a co-worker.
    - Divide the load into smaller parts, even if it means making a few extra trips.
    - Use a lifting device, such as a dolly.
2. **Bend your knees.** This is the single, most important rule that applies to lifting. Take a tip from professional weight lifters. They can lift tremendous weights because they lift with their legs, not their backs.
    - When lifting a box, for example, position your feet close to it.
    - Center yourself over the box.

- Bend your knees and get a good grip on the box.

When lifting objects—and setting them down—bend your knees, and let your legs (not your back) do the work.

- Lift straight up, slowly and smoothly.
- Allow your legs, not your back, to do the work.

3. **Do not twist or turn your body once you have made the lift.** Keep the load close to your body, and keep it steady. Any sudden twisting or turning could result in taking out your back. When you do need to turn, continue to hold the load near your body, and turn your body by changing (turning) your foot positions.

4. **Make sure you can carry the load to its destination before attempting to move it.** Also, make sure your path is clear of obstacles and that there are no hazards, including dropped food or spilled beverages. Have sure footing before setting out.

5. **Set down the load properly.** Setting the load down is just as important as lifting it. Lower the load slowly by bending your knees and letting your legs do most of the work. Don't let go of the load until it is securely in its place.

6. **Push, don't pull.** Always push the object when possible. For example, if you move an object on rollers, pushing puts less strain on your back.

Always push, never pull.

Wrap-up

A number of factors can cause backache, but there are things you can do to prevent it. Every time you need to lift something—whether at work or at home—remind yourself to use good lifting techniques, no matter how much the item may weigh. Remember, a specific weight lifted may actually have little to do with any single injury. Quite possibly, the harm is due to repeated incorrect lifting techniques.

Employee ____
Instructo____
Date ____
Company____

REVIEW OF LIFTING TECHNIQUES TO SAVE YOUR BACK

1. Poor posture—whether standing, sitting, or lying down—can cause back pain.
   a. True
   b. False

2. To avoid backache, you should sleep:
   a. On your stomach.
   b. In any position that is comfortable.
   c. On your side with knees bent, or on your back.
   d. On the floor.

3. If you are overweight, and carry much of the extra weight in your stomach:
   a. It should not cause much of a problem for your back.
   b. You are putting extra strain on your back.
   c. It will lead to poor posture.
   d. None of the above.

4. You can avoid back injury by:
   a. Practicing good posture.
   b. Eating a proper diet and exercising.
   c. Not putting minor strains on your back over and over again.
   d. All of the above.

5. Stress is normal and no amount of it can affect your back.
   a. True
   b. False

6. If you are standing for long periods of time, you can ease the strain on your back by:
   a. Keeping your weight evenly balanced on both feet.
   b. Not leaning to one side.
   c. Placing one foot on a ledge and switching foot positions often.
   d. All of the above.

7. Which is not a good lifting technique?
   a. Test the weight of the load by lifting at one of the corners.
   b. Bend your knees before lifting and let your legs do the work.
   c. As you lift the load, twist at the waist to face the proper direction.
   d. Set the load down by bending your knees and letting your legs do the work.

8. If you cannot lift a load by yourself, you can:
   a. Use a dolly.
   b. Ask for help from a co-worker.
   c. Divide the load into smaller parts and make a few trips.
   d. All of the above.

9. Setting a load down properly is just as important as lifting it properly.
   a. True
   b. False

10. Whenever possible, to put less strain on your back, you should move a load by:
    a. Pulling
    b. Pushing

GLOSSARY OF FOOD SAFETY-RELATED TERMS

Allergen—A food, such as peanuts, that causes an allergic reaction.

Antibacterial—Directed at or effective against bacteria. Antibacterial soap eliminates or kills bacteria.

Bacteria—Minute, one-celled microorganisms that can contaminate food and cause illness, or even death, in people.

Biological hazard—In food safety, this term refers to microorganisms, like bacteria, in food that pose a risk of causing illness in people.

Botulism—A severe, sometimes fatal food poisoning caused by eating a toxin produced from improperly canned or preserved food.

Chemical hazard—A hazard posed to food by chemicals that might accidentally get into the food.

Cleaning solution—A chemical, such as bleach or detergent, that is used with or without water to clean surfaces.

Contaminant—Anything that can get into food that is not supposed to be there, such as glass, metal, bacteria, hair, cleansers, and jewelry.

Contamination—The presence of extraneous, especially infectious, material that makes food impure or harmful.

Critical control point (CCP)—A step or procedure in the food handling process where a food safety hazard can be controlled.

Cross contamination—The process during which the material that does not belong in food, such as bacteria, moves from one place to another.

Dehydration—Excessive loss of water and other body fluids that occurs during illness.

Deviation—A condition outside of the critical limits set for food.

***E. coli* (*Escherichia coliform*)**—Bacteria found in the gastrointestinal tract that can cause diarrheal diseases.

FIFO—An acronym that stands for "first in, first out." In foodservice, this means use food in the order it was received.

Flow of food—The progressive steps food follows, from receiving, to storing, to serving.

Gravity—The force that pulls things toward the Earth.

HACCP—Acronym for Hazard Analysis and Critical Control Points. A system aimed toward creating safe food by correcting risks before they occur.

Hazard—A source of danger. In foodservice, a hazard can be physical, chemical, or biological.

Hazard analysis—The first part of a HACCP program, which identifies every way food can become unsafe to eat.

Hepatitis—Inflammation of the liver, caused by infection or toxins, with symptoms of jaundice, fever, liver enlargement, and abdominal pain.

Hepatitis A—A form of hepatitis caused by a virus that is transmitted by eating infected food or water.

Hepatitis B and C—Forms of hepatitis caused by a virus that is transmitted by infected blood or by contaminated instruments. The disease may become severe or chronic, causing serious liver damage.

HIV—Human immunodeficiency virus that is the cause of acquired immunodeficiency syndrome (AIDS).

Hygiene—The practices, including cleanliness, that serve to promote and preserve health.

Infect—To contaminate with disease-producing matter.

Irradiation—Application of gamma rays or other radiation to prevent spoilage of food.

Listeriosis—A severe bacterial disease affecting animals and occasionally humans, characterized by sudden fever, nausea, delirium, and coma.

Microorganisms—Living things that are so small, they can only be seen with the aid of a microscope.

Molds—Various fungi (generally a circular colony with a woolly or furry appearance) that grow on food and contribute to its spoilage.

Momentum—A property that a moving body has due to its mass and motion.

Parasite—A plant or animal living in or on another organism, usually to its harm.

Pathogen—An agent that causes disease, especially a living microorganism, such as a bacterium, virus, or fungus.

Pesticide—An agent used to destroy pests.

pH—A measure of the acidity or alkalinity of a solution.

Physical hazard—Any object that can get into food and contaminate it, such as fingernails, bits of glass, metal, hair, and packaging material.

Salmonella—Any of various bacteria, many of which are pathogenic, causing food poisoning, typhoid, and other infectious diseases.

Sanitary—Free from elements, such as filth or pathogens, that endanger health; hygienic.

Sanitation—The act or process of making sanitary, particularly to protect public health.

Sanitizer—An agent that, when applied properly, makes a surface sanitary.

***Staph* (*Staphylococcus aureus*)**—Any of various bacteria causing boils and other infections.

Temperature danger zone—40°–140°F (or 4°–60°C). The temperature range that accelerates growth of harmful bacteria and molds in food.

Time and temperature controls—The procedures designed to keep foods out of the temperature danger zone.

Toxin—A poisonous substance that is produced by living organisms and that is capable of causing disease.

Traction—The adhesive friction (grip) of a body on a surface on which it moves.

Unpasteurized—Not subject to pasteurization (a process of heating a beverage, such as milk, to a specific temperature for a specific period of time in order to kill microorganisms that could cause disease or spoilage).

Verification—The final principle of a HACCP program that confirms the system is working properly.

Virus—Any of a large group of submicroscopic infectious agents that cause disease.

NOTES

NOTES

NOTES

NOTES

NOTES

NOTES

NOTES

NOTES

NOTES

NOTES

NOTES

NOTES